PYTHON FOR DATA SCIENCE

STEP-BY-STEP PYTHON CRASH COURSE ON HOW TO COME UP EASILY WITH YOUR FIRST DATA SCIENCE PROJECT FROM SCRATCH IN LESS THAN 7 DAYS. INCLUDES PRACTICAL EXERCISES

TED WOLF

Copyright - 2020 - Ted Wolf

All rights reserved.

The content contained within this book may not be reproduced, duplicated or transmitted without direct written permission from the author or the publisher.

Under no circumstances will any blame or legal responsibility be held against the publisher, or author, for any damages, reparation, or monetary loss due to the information contained within this book. Either directly or indirectly.

Legal Notice:

This book is copyright protected. This book is only for personal use. You cannot amend, distribute, sell, use, quote or paraphrase any part, or the content within this book, without the consent of the author or publisher.

Disclaimer Notice:

Please note the information contained within this document is for educational and entertainment purposes only. All effort has been executed to present accurate, up to date, and reliable, complete information. No warranties of any kind are declared or implied. Readers acknowledge that the author is not engaging in the rendering of legal, financial, medical or professional advice. The content within this book has been derived from various sources. Please consult a licensed professional before attempting any techniques outlined in this book.

By reading this document, the reader agrees that under no circumstances is the author responsible for any losses, direct or indirect, which are incurred as a result of the use of information contained within this document, including, but not limited to, - errors, omissions, or inaccuracies.

TABLE OF CONTENTS

INTRODUCTION		5
CHAPTER - 1	BASICS OF PYTHON FOR DATA SCIENCE	11
CHAPTER - 2	SETTING UP PYTHON ENVIRONMENT FOR DATA SCIENCE	19
CHAPTER - 3	MACHINE LEARNING WITH SCIKIT-LEARN AND HOW IT FITS WITH DATA SCIENCE	27
CHAPTER - 4	APPLICATION OF MACHINE LEARNING USING SCIKIT-LEARN LIBRARY	39
CHAPTER - 5	DATA STRUCTURES	49
CHAPTER - 6	DATA SCIENCE ALGORITHMS AND MODELS	63
CHAPTER - 7	DATA AGGREGATION AND GROUP OPERATIONS	75
CHAPTER - 8	PRACTICAL CODES AND EXERCISES TO USE PYTHON	81
CHAPTER - 9	FUNCTIONS AND MODULES IN PYTHON	95
CHAPTER - 10	DATA SCIENCE AND THE CLOUD	107
CHAPTER - 11	DATA MINING	115
CONCLUSION		125

```
         = -1;       return c;
    sh(a[c]);    }
    /\n|\r)/gm, " "), b
   p_array.length;
  c.push(inp_array[a]),
   , inp_array));
    " ");      -1 <  b
);           -1 < b
  on use_array(a, b) {
    y(a, b) {      for (var
                   var c = -1,
```

INTRODUCTION

When data science first came about, it was only in the hands of the scientists and a few daring accountants. This is very understandable because without truly getting a glimpse of what data science consists of, it is easy to assume that it is for the strong at heart and those who enjoy solving 'boring' problems. But it is 2020, and the data science rave is everywhere. This can only mean one thing—that it is very important to the world we currently live in. More and more data scientists are in demand; every day, more and more technologies are built to help the concept of data science. But what exactly is it that makes data science that important to the 21st-century person or organization? Simple—DATA.

People, organizations, and countries need data, no matter the level they are in. Statistics are needed to gauge development, make progress reports, and a whole lot more wherever it is that we turn to. This begs the question—what is data science? Although it may not have a particular definition that is generally acceptable, because it has become a global phenomenon, it is an interdisciplinary subject that comprises of three distinct yet overlapping areas. These areas are statistics, computer science, and domain expertise.

A data scientist will then be an expert who can model and summarize data sets, design, and use algorithms to effectively store, process and visualize the data gotten while being able to form the right questions and putting the answers into context. The reality, though, is that the best data scientists today work in teams. Because of the variety of skills that are needed in the field, it's rare to find those who have perfected all the skill sets. So, if you're looking to learn data science, it's fine not to have all the requisite skills yet. A perfect summary of a data scientist's portfolio will be—data capturing, data analysis, and data presentation. Data science entails the use of furthered mathematical techniques, statistics, and big data.

Data scientists aren't just the 'boring' people who'd just sit before their computers crunching numbers; they're those who can answer questions with even more questions, help us make better decisions with the information we have, create suggestions for options based on preceding choices, make robots see objects, and a whole lot more. In fact, data science is found in literally any concept, ideology, or industry that there is today (you name it) that we can hardly look anywhere without feeling its effects. Data science helps in sharing the bewildering experiences we get from technology today. When we say data science is what helps us understand and accept what we regard as our reality today, it's really nothing but the truth.

Although the concept of Data Science (the process of quantifying and understanding statistics) is relatively new, the principles and mathematics behind it have always existed. So, it'd be great to approach data science not as an entirely new domain of knowledge, but as a path through which you can apply the knowledge you already have.

You may not know this, but in one way or the other, you've applied data science to one or two of your daily activities. Take, for instance, when you use search engines to look for something. At some point, it'll make suggestions on some alternatives for you. Those alternative terms are gained through data science. When a doctor makes a prognosis, one way he could have known that your lump isn't cancerous was through data science.

This book, in the first place, not only intends to give a simple and useful introduction to data science but also to show you how important data science is to our everyday lives. You not only know how to answer the questions brought forward, but you're also sure where they can be employed. So whatever field you find yourself in, whether you're predicting stock returns, optimizing online ad clicks, reporting election results, or whatever field it is that data is required (which is everywhere), you'll be able to stand out with better knowledge and know-how of data science.

This book is made to be a tool that, first off, harmonizes data science with Python. It looks at connecting the dots between these two interrelated computer concepts and making them one. It will highlight for you a million and more reasons why learning data science with Python is one of the best ways to go about it and why you should take advantage of what it brings.

Python was first implemented in 1989 and is regarded as highly user-friendly and simple to learn programming language for entry-level coders and amateurs. It is a high-level programming language, commonly used for general purposes. It was originally developed by Guido van Rossum at the "Center Wiskunde & Informatica (CWI), Netherlands," in the 1980s and introduced by the "Python Software Foundation" in 1991. It is considered ideal for people who are new to programming or coding and need to understand the basics of programming. This is due to the fact that Python reads almost the same as English. Therefore, less time is needed to understand how the language works, and the focus may be on learning the basics of programming.

Python is an interpreted language that supports automatic memory management and object-oriented programming. This highly intuitive and flexible programming language can be used to code programs such as machine learning algorithms, web applications, data mining and visualization, game development.

CHAPTER - 1

BASICS OF PYTHON FOR DATA SCIENCE

What Is Data Science?

Data science is a gathering of different instruments, data interfaces, and calculations with AI standards (algorithms) to find concealed patterns from raw data. This data is put away in big business data distribution warehouses and utilized in inventive approaches to create business value.

A data examiner (analyst) and a data scientist are unique. An analyst attempts to process data history and clarify what is happening. In contrast, a data researcher needs different propelled calculations of AI (algorithms of machine learning) for an event of a specific occasion by utilizing analysis.

Python and Its History

Python is a globally useful, high-quality, translated programming language. Developed by Guido van Rossum and first released in 1991, the Python

Foundation emphasizes code clarity by making the most of critical space. Its language is developed and designed with object methodology to allow software engineers to compose clear and logical code for small and large-scale projects.

Python was first developed in the late 1980s as the successor to the ABC language. Python 2.0, released in 2000, featured snapshots, such as degradation concepts and a garbage collection framework, suitable for collecting reference cycles. Python 3.0, downloaded in 2008, was a notable language modification, and much of Python 2's code does not run unmodified in Python 3. Language designer Guido van Rossum was solely tasked with committing until July 2018, but he now shares management as one person on a five-member board.

Unique Features and Philosophy

Python is a versatile programming language that supports Object-Oriented Programming (OOP) and other practical computer program languages. Initially, it was not designed for data science, but as a field, professionals started using it for data analysis, and it became a priority for data science. Many different standards are bolstered utilizing expansions, including a plan by contract and rationale programming. Likewise, it includes dynamic name goals (late authoritative), which tie technique and variable names during system operations. The standard library has two modules that actualize useful devices acquired from Haskell and Standard ML.

Unlike incorporating most of its utility into its core, Python was meant to be deeply scalable.

Python is moving toward less complex and less mixed scoring and structure while allowing engineers to make decisions on their approach to coding. Contrary to the saying, "Perl there is more than one approach," Python understands that "there must be one, and ideally one, clear approach to doing this." Alex Martelli of the Python Software Foundation and author of the book Python says that "portraying something as 'sharp' is not a compliment to Python culture."

Python engineers have tried to keep a strategic distance from initial progress and lock in patches in unnecessary parts of CPython that will offer minimal speed increases at the expense of clarity. When speed is important, a Python software engineer can transfer basic sync capabilities to extensions written in dialects. For example, C, or use PyPy, one in the time compiler name. Cython is also accessible, which interprets Python content in C and makes direct C-level API calls to the Python translator.

Python's progress has been vastly improved with the Python Enhancement Proposal (PEP) process. This included collecting community feedback on issues and recording decisions about the Python framework. The Python coding style is included in PEP 8. Excellent PEPs are rated and evaluated by the Python community and the Python dashboard.

Language improvement is compared to progress in using CPython reports. The mailing list, Python-dev, is an essential discussion on the evolution of the language. Specific issues are discussed in the Roundup debugger maintained at Python.org. Development was initially carried out on a self-supplied Mercurial source code repository, until Python was moved to GitHub in January 2017.

CPython open discards are available in three types, which determine how much of the customization number is incremented.

Backward variations are where the code is required to break and should transfer naturally. The initial part of the configuration number increases. These vaccines are rare. For example, Custom 3.0 was downloaded eight years after 2.0.

Large or "standard" shots look like a clock and include new features. The second part of the form number increases. All major variants support bug fixes long after they are released.

Non-new error correction rejections occur at regular intervals and occur when a sufficient number of upstream errors have been corrected since the last discharge. Security vulnerabilities are also defined in these discards—the third and last part of the form number increases.

Many alpha and beta downloads are also downloaded as a peek and for testing before final downloads. Although there is an unpleasant schedule for each exemption, it is often postponed if the password is not ready. The Python progress team checks the status of the code by running a huge set of unit tests during the upgrade and using the uninterruptible BuildBot join system. The Python engineering community has also contributed more than 86,000 programming modules. The real school conference for Python is PyCon. There are also excellent Python training programs, for example, Pyladies.

Python Applications

Python is known for its broadly useful nature that makes it relevant in practically every space of programming advancement. Python can be used in a plethora of ways for improvement; there are specifying application territories where Python can be applied.

Web-Applications

We can utilize Python to create web applications. It gives libraries to deal with web conventions, for example, HTML and XML, JSON, email handling, demand, beautiful soup, Feedparser, and so on. Additionally, there are Frameworks. For example,

Django, Pyramid, Flask, and so on to structure and develop electronic applications. Some significant improvements are PythonWikiEngines, PythonBlogSoftware, and so on.

Desktop GUI Applications

Python gives a Tk-GUI library to create UI in Python-based application. Another valuable toolbox includes wxWidgets, Kivy, and is useable on a few stages. The Kivy is well known for comp sing multitouch applications.

Software Development

Python is useful for programming advanced processes. It functions as a help language and can be utilized for fabricating control and the board, testing, and so forth.

Scientific and Numeric

Python is mainstream and generally utilized in logical and numeric figuring. Some helpful libraries and bundles are SciPy, Pandas, IPython, and so forth. SciPy is a library used for the collection of bundles of designing, science, and arithmetic.

Business Applications

Python is utilized to manufacture business applications, like ERP and online business frameworks. Tryton is an abnormal state application stage.

Console Based Application

It can be utilized for support-based applications. For instance: IPython.

Audio or Video-based Applications

Python is great for playing out various assignments and can be utilized to create media applications. Some of the authentic applications are cplay, TimPlayer, and so on.

Enterprise Applications

Python can be utilized to make applications that can be utilized inside an Enterprise or an Organization. Some ongoing applications are Tryton, OpenERP, Picalo, etc.

Applications for Images

Utilizing Python, a few applications can be created for a picture. Various applications include VPython, Gogh, and imgSeek.

Why Python to Conduct Data Analysis

Different programming languages can be utilized for data science (for example, SQL, Java, Matlab, SAS, R, and some more), yet Python is the most favored by data researchers among the various programming languages in this rundown. Python has some exceptional features, including:

- Python is solid and basic with the goal that it is anything but difficult to gain proficiency in the language. You don't have to stress over its linguistic structure on the off chance that you are an amateur. Its syntax is similar to English writing; that's why it is an easy to use programming language.

- Python supports almost all platforms, like Windows, Mac, and Linux.
- It has multiple data structures with which complex calculations can easily be simplified.
- Python is an open-source programming language that enables the data scientists to get pre-defined libraries and codes to perform their tasks.
- Python can perform data visualization, data investigation, and data control.
- Python serves different ground-breaking libraries for algorithms and logical calculations. Different complex logical figuring and AI calculations can be performed utilizing this language effectively in a moderately basic sentence structure.

CHAPTER - 2
SETTING UP PYTHON ENVIRONMENT FOR DATA SCIENCE

Now that you have decided to work with the Python code to help you do some of your own programming and coding, it is time to get to work with installing this on your computer. You will not be able to get all of the work done or use any of the coding languages if all of the different parts and files that come with the Python code are not on your computer. We will take some time to look at how you can install Python on your computer, no matter which operating system you want to work with.

For these steps, we are going to assume that you are getting Python from www.python.org. There are other resources where you are able to get this programming language, but this one is often the easiest because it is going to have all of the files and extensions that you need to make Python work, and all of them are free to use. Other sources can provide some more features and other things that you need, but they do not always work well, or they may be missing some of the parts that you need. So, let's dive in and see how we can get the folders of Python set up with our computer.

Installing On the Mac OS X

The first option that we are going to look at when we want to add on Python to our operating system is the Mac OS X. This is a popular option when it comes to the operating system on a computer, and it is going to work just fine with some of the codings that we decide to do with Python. However, you need to double-check the system because some of the Python versions are going to automatically be included in this operating system. To see which version of the Python program is found on your system, it will include the following code:

Python – V

This is going to show you the version you get, so a number will come up. You can also choose to install Python 3 on this system if you would like, and it isn't required to uninstall the 2.X version on the computer. To check for the 3.X installation, you just need to open up the terminal app and then type in the following prompt:

Python3 – V

The default on OS X is that Python 3 is not going to be installed at all. If you want to use Python 3, you can install it using some of the installers that are on Python.org. This is a good place to go because it will install everything that you need to write and execute your codes with Python. It will have the Python shell, the IDLE development tools, and the interpreter. Unlike what happens with Python 2.X, these tools are installed as a standard application in the Applications folder.

Being able to run the IDLE and the Python shell is going to be dependent on which version you choose and some of your own personal preferences. You can use the following commands to help you start the shell and IDLE applications:

- For Python 2.X just type in "Idle"
- For Python 3.X, just type in "Idle3"

Installing on a Windows System

A Windows system is also able to work with Python. There is not going to be a version of this program on Windows, though, simply because the Windows system has its own programming language that you are able to work with. This does not mean that you are limited to only using that language, but it does mean that you are going to have to spend some time installing Python and picking out the version of Python that you would like to use.

There are a few steps that you are able to use in order to make sure that you can install the Python program on your system. It can look a bit intimidating when you first get started, but you will find that these steps only take a few minutes, so it isn't as bad as it seems. Some of the steps that you need to follow in order to make sure that Python is installed properly on your Windows computer will include:

1. To start this, it is time to visit the official download page for Python and then grab the Windows Installer. You are able to choose which version of Python that you would like to work with. Your default of this is going to give you the 32-bit version of the language, but you can go through and click on the 64-bit version if this is what you need for your computer system.

2. Now you can go through and push on the right-click button on the installer and then allow it to run as Administrator. There is going to be a point that gives you two options. You will want to go with the button that allows for customizing the installation.

3. On the following screen, you need to make sure that you look under the part for Optional Features and click on all of the boxes that are there. When those boxes are filled out, you can click the Next button.

4. While you are still in the part for Advanced Options, you can pick out the location where you would like to install Python. Click on Install and then wait a few minutes in order to get this installation finished. Then close out of the installer when it is all done.

5. At this point, you can set up the PATH variable to the system to make sure that it is going to include all of the directories that come with the packages and to make sure that all of the other components that you need to show up as well. The steps that you can follow to make all of this happen includes:

a. Start this part by opening up your Control Panel. This is easily done by clicking on your taskbar and then typing in "Control Panel." Click on the icon that shows up.

b. While you are in this part of the process, it is time to do a search for the Environment. Then you can click on the part that allows you to edit the System Environment Variables. From here, you can then click on the button that allows you to enter into Environment Variables.

c. Head over to the section that is for User Variables. You can choose to edit the PATH variable that is already there for your use, or you can decide to create a brand new one.

d. If you see that this system doesn't provide you with a variable for the PATH, then it is time to create your own. You can create this by clicking on NEW. Make the name of the PATH variable and add it to the directories that you want. Close all the Control Panel dialogs and then click on Next to finish it up.

6. At this point, you are able to open up the command prompt on your Windows computer. This is done by clicking on your Start Menu and then going to the Windows System and then into Command Prompt. Type in the word "python." This will make sure that the interpreter of Python is going to load up for you.

At this point, the program is going to be set up and ready to use on your Windows system. You can choose to open up the other parts of the system as well to make sure that all of the parts that you need are in one place, and then it is time to write out any code that you want when the time is right.

How to Install Python on Your Linux System?

Now that we have been able to explore how to install Python on your Windows computer and your Mac OS X, it is time to move on to some of the steps that you can use to get this language installed on your Linux system as well. There are many individuals and programmers alike who are moving over to the Linux system, so it is one that we need to spend some time learning how to use for our needs.

24

The first thing to do here is to see if there is a version of Python 3 already on your system. You can open up the command prompt on Linux and then run the following code:

$ Python3 - - version

If you are on Ubuntu 16.10 or newer, then it is a simple process to install Python 3.6. You just need to use the following commands:

$ Sudo apt-get update

$ sudo apt-get install Python3.6

If you are relying on an older version of Ubuntu or another version, then you may want to work with the deadsnakes PPA, or another tool, to help you download the Python 3.6 version. The code that you need to do this includes:

$ Sudo apt-get install software-properties-common

$ Sudo add-apt repository ppa: deadsnakes/ppa

Suoda apt-get update

$ Sudo apt-get install python3.6

The thing to remember here is that if you have spent some time working with a few of the other distributions that come with Linux, it is likely that your system is going to have a version of Python 3 already installed on there. If you do not see this here, you can choose to use the package manager of the distribution. Or you can go through the steps that we have above in order to help you to install any version of Python that you want before using the program.

Understanding the Interpreter in Python

The standard installation of Python, when you do it on python.org, is going to contain documentation, information on licensing, and three main files to execute that help you develop and then run the scripts that you run on python. These include the Python interpreter, IDLE, and Shell.

First is the Python interpreter. This is important because it is responsible for executing the scripts that you decide to write. The interpreter can convert the .py script files into instructions and then processes them according to the type of code that you write in the file.

Then there is the Python IDLE. This is known as the integrated development and learning environment. It is going to contain all of the tools that are needed to develop your programs in Python. You will find tools for debugging, the text editor, and the shell with this. Depending on the version of Python that you choose, the IDLE can either be extensive or pretty basic. You can also pick out your own IDLE if there is another version that you like better. Many people like to find new text editors because they think the one with Python doesn't have the right features, but the one from Python is just fine for the codes we will do, so it's not necessary to pick a different one.

CHAPTER - 3

MACHINE LEARNING WITH SCIKIT-LEARN AND HOW IT FITS WITH DATA SCIENCE

Scikit-Learn is a versatile Python library that is useful when building data science projects. This powerful library allows you to incorporate data analysis and data mining to build some of the most amazing models. It is predominantly a machine learning library, but can also meet your data science needs. There are many reasons why different programmers and researchers prefer Scikit-Learn. Given the thorough documentation available online, there is a lot that you can learn about Scikit-Learn, which will make your work much easier, even if you don't have prior experience. Leaving nothing to chance, the API is efficient, and the library is one of the most consistent and uncluttered Python libraries you will come across in data science.

Like many prolific Python libraries, Scikit-Learn is an open-source project. There are several tools available in Scikit-Learn that will help you perform data mining and analysis assignments easily. Earlier in the book, we mentioned that some Python libraries would cut across different dimensions. This is one of them. When learning about the core Python libraries, it is always important that you understand you can implement them across different dimensions.

Scikit-Learn is built on Matplotlib, SciPy, and NumPy. Therefore, knowledge of these independent libraries will help you get an easier experience using Scikit-Learn.

Uses of Scikit-Learn

How does Scikit-Learn help your data analysis course? Data analysis and machine learning are intertwined. Through Scikit-Learn, you can implement data into your machine learning projects in the following ways:

Classification

Classification tools are some of the basic tools in data analysis and machine learning. Through these tools, you are able to determine the appropriate category necessary for your data, especially for machine learning projects. A good example of where classification models are used is in separating spam emails from legitimate emails.

Using Scikit-Learn, some classification algorithms you will come across include random forest, nearest neighbors, and support vector machines.

Regression

Regression techniques in Scikit-Learn require that you create models that will autonomously identify the relationships between input data and output. From these tools, it is possible to make accurate predictions, and perhaps we can see the finest illustration of this approach in the financial markets or the stock exchanges. Common regression algorithms used in Scikit-Learn include Lasso, ridge regression, and support vector machines.

Clustering

Clustering is a machine learning approach where models independently create groups of data using similar characteristics. By using clusters, you can create several groups of data from a wide dataset. Many organizations access customer data from different regions. Using clustering algorithms, this data can then be clustered according to regions. Some of the important algorithms you should learn include mean-shift, spectral clustering, and K-means.

Model selection

In model selection, we use different tools to analyze, validate, compare, and contrast, and finally choose the ideal conditions that our data analysis projects will use in operation. For these modules to be effective, we can further enhance their accuracy using parameter tuning approaches like metrics, cross-validation, and grid search protocols.

Dimensionality reduction

In their raw form, many datasets contain a high number of random variables. This creates a huge problem for analytics purposes. Through dimensionality reduction, it is possible to reduce the challenges expected when having such variables in the dataset. If, for example, you are working on data visualizations and need to ensure that the outcome is efficient, a good alternative would be eliminating outliers altogether. To do this, some techniques you might employ in Scikit-Learn include non-negative matrix factorization, principal component analysis, and feature selection.

Preprocessing

In data science, preprocessing tools used in Scikit-Learn help you extract unique features from large sets of data. These tools also help in normalization. For instance, these tools are helpful when you need to obtain unique features from input data like texts, and use the features for analytical purposes.

Representing Data in Scikit-Learn

If you are working either individually or as a team on a machine learning model, working knowledge of Scikit-Learn will help you create effective models. Before you start working on any machine learning project, it is mandatory that you take a refresher course on data representation. This is important so that you can present data in a manner such that your computers or models will comprehend easily. Remember that the kind of data you feed the computer will affect the outcome. Scikit-Learn is best used with tabular data.

Tabular Data

The tables are simple two-dimensional representations of some data. Rows in a table identify the unique features of each element within the data set. Columns, on the other hand, represent the quantities or qualities of the elements you want to analyze from the data set. In our illustration for this section, we will use the famous Iris dataset. Lucky for you, Scikit-Learn comes with the Iris dataset loaded in its library, so you don't need to use external links to upload it. You will import this dataset into your programming environment using the Seaborn library as a DataFrame in Pandas.

The Iris dataset comes preloaded into Scikit-Learn, so pulling it to your interface should not be a problem. When you are done, the output should give you a table whose columns include the following:

- sepal_length
- sepal_width
- petal_length
- petal_width
- species

We can deduce a lot of information from this output. Every row represents an individual flower under observation. In this dataset, the number of rows infers the total number of flowers present in the Iris dataset. In Scikit-Learn, we will not use the term rows, but instead, refer to them as samples. Based on this assertion, it follows that the number of rows in the Iris dataset is identified as n_samples.

On the same note, columns in the Iris dataset above provide quantitative information about each of the rows (samples). Columns, in Scikit-Learn, are identified as features; hence the total number of columns in the Iris dataset will be identified as n_features.

What we have done so far is to provide the simplest explanation of a Scikit-learn table using the Iris dataset.

Features Matrix

From the data we obtained from the Iris dataset, we can interpret our records as a matrix or a two-dimensional array. If we choose to use the matrix, what we have is a features matrix.

By default, features matrices in Scikit-Learn are stored in variables identified as x. Using the data from the table above to create a features matrix, we will have a two-dimensional matrix that assumes the following shape [n_samples, n_features]. Since we are introducing arrays, this matrix will, in most cases, be part of an array in NumPy. Alternatively, you can also use Pandas DataFrames to represent the features matrix.

Rows in Scikit-Learn (samples) allude to singular objects that are contained within the dataset under observation. If, for example, we are dealing with data about flowers as per the Iris dataset, our sample must be about flowers. If you are dealing with students, the samples will have to be individual students. Samples refer to any object under observation that can be quantified in measurement.

Columns in Scikit-Learn (features) allude to unique descriptive observations we use to quantify samples. These observations must be quantitative in nature. The values used in features must be real values, though in some cases, you might come across data with discrete or Boolean values.

33

Target Arrays

Now that we understand what the features matrix (x) is, and its composition, we can take a step further and look at target arrays. Target arrays are also referred to as labels in Scikit-Learn. By default, they are identified as (y).

One of the distinct features of target arrays is that they must be one-dimensional. The length of a target array is n_samples. You will find target arrays either in the Pandas series or in NumPy arrays. A target array must always have discrete labels or classes, and the values must be continuous if using numerical values. For a start, it is wise to learn how to work with one-dimensional target arrays. However, this should not limit your imagination. As you advance into data analysis with Scikit-Learn, you will come across advanced estimators that can support more than one target array. This is represented as a two-dimensional array, in the form [n_samples, n_targets].

Remember that there exists a clear distinction between target arrays and features columns. To help you understand the difference, take note that target arrays identify the quantity we need to observe from the dataset. From our knowledge of statistics, target arrays would be our dependent variables. For example, if you build a data model from the Iris dataset that can use the measurements to identify the flower species, the target array in this model would be the species column.

The diagrams below give you a better distinction between the target vector and the features matrix:

Diagram of a Target vector

Diagram of a Features Matrix

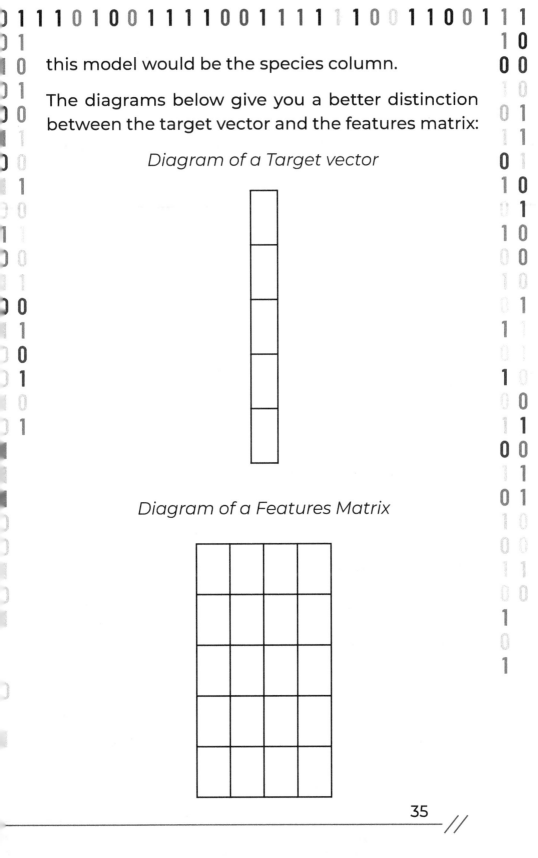

Understanding the API

Before you start using Scikit-Learn, you should take time and learn about the API. According to the Scikit-Learn API paper, the following principles are the foundation of the Scikit-Learn API:

Inspection

You must show all the parameter values in use as public attributes

Consistency

You should use a limited number of methods for your objects. This way, all objects used must have a common interface, and to make your work easier, ensure the documentation is simple and consistent across the board.

Limited object hierarchy

The only algorithms that should use basic Python strings are those that belong to Python classes.

Sensible defaults

For models that need specific parameters unique to their use, the Scikit-Learn library will automatically define the default values applicable

Composition

Given the nature of machine learning assignments, most of the tasks you perform will be represented as sequences, especially in relation to the major machine learning algorithms.

Why is it important to understand these principles? They are the foundation upon which Scikit-Learn is built; hence they make it easier for you to use this Python library. All the algorithms you use in Scikit-Learn, especially machine learning algorithms, use the estimator API for implementation. Because of this API, you can enjoy the consistency in development for different machine learning applications.

```
ters. and p.name
meters.contains(
    and p.age =
dQuery<Person> query
parameters.contains(
uery.setParameter(
```

CHAPTER - 4
APPLICATION OF MACHINE LEARNING USING SCIKIT-LEARN LIBRARY

To understand how the Scikit-Learn library is used in the development of a machine learning algorithm, let us use the "Sales_Win_Loss data set from IBM's Watson repository" containing data obtained from the sales campaign of a wholesale supplier of automotive parts. We will build a machine learning model to predict which sales campaign will be a winner and which will incur a loss.

The data set can be imported using Pandas and explored using Pandas techniques such as "head (), tail (), and dtypes ()." The plotting techniques from "Seaborn" will be used to visualize the data. To process the data Scikit-Learn's "preprocessing.LabelEncoder ()" will be used and "train_test_split ()" to divide the data set into a training subset and testing subset.

To generate predictions from our data set, three different algorithms will be used, namely, "Linear Support Vector Classification and K-nearest neighbor's classifier." To compare the performances of these algorithms Scikit-Learn library technique "accuracy_score" will be used. The performance score of the models can be visualized using Scikit-Learn and "Yellowbrick" visualization.

Importing the Data Set

To import the "Sales_Win_Loss data set from IBM's Watson repository," the first step is importing the "Pandas" module using "import pandas as pd."

Then we leverage a variable url as: "https://community.watsonanalytics.com/wpcontent/uploads/2015/04/ WA_Fn-UseC_-Sales-Win-Loss.csv" to store the URL from which the data set will be downloaded.

Now, "read_csv () as sales_data = pd. read_csv(url)" technique will be used to read the above "csv or comma separated values" file, which is supplied by the Pandas module. The csv file will then be converted into a Pandas data framework, with the return variable as "sales_data," where the framework will be stored.

For new 'Pandas' users, the "pd. read csv ()" technique in the code mentioned above will generate a tabular data structure called "data framework," where an index for each row is contained in the first column, and the label/name for each column in the first row are the initial column names acquired from the data set. In the above code snippet, the "sales data" variable results in a table depicted in the picture below.

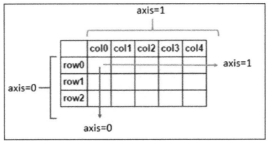

In the diagram above, the "row0, row1, row2" represent individual record index, and the "col0, col1, col2" represent the names for individual columns or features of the data set.

With this step, you have successfully stored a copy of the data set and transformed it into a "Pandas" framework!

Now, using the "head () as Sales_data. Head ()" technique, the records from the data framework can be displayed as shown below to get a "feel" of the information contained in the data set.

41

	opportunity number	supplies subgroup	supplies group	region	route to market	elapsed days in sales stage	opportunity result
0	1641984	Exterior Accessories	Car Accessories	Northwest	Fields Sales	76	Won
1	1658010	Exterior Accessories	Car Accessories	Pacific	Reseller	63	Loss
2	1674737	Motorcycle Parts	Performance & Non-auto	Pacific	Reseller	24	Won
3	1675224	Shelters & RV	Performance & Non-auto	Midwest	Reseller	16	Loss

Data Exploration

Now that we have our own copy of the data set, which has been transformed into a "Pandas" data frame, we can quickly explore the data to understand what information can tell, can be gathered from it, and accordingly to plan a course of action.

In any ML project, data exploration tends to be a very critical phase. Even a fast data set exploration can offer us significant information that could be easily missed otherwise, and this information can propose significant questions that we can then attempt to answer using our project.

Some third-party Python libraries will be used here to assist us with the processing of the data so that we can efficiently use this data with the powerful algorithms of Scikit-Learn. The same "head ()" technique that we used to see some initial records of the imported data set in the earlier section can be used here. As a matter of fact, "(head)" is effectively capable of doing much more than displaying data records and customize the "head ()" technique to display only selected records with commands like "sales_data.head(n=2)." This command will selectively display the first two records of the data set. At a quick glance, it's obvious that columns such as "Region" and "Supplies Group" contain string data, while columns such as "Opportunity Result," "Opportunity Number," etc. are comprised of integer values. It can also be seen that there are unique identifiers for each record in the 'Opportunity Number' column.

Similarly, to display select records from the bottom of the table, the "tail() as sales_data.tail()" can be used.

To view the different data types available in the data set, the Pandas technique "dtypes() as sales_data.dtypes" can be used. With this information, the data columns available in the data framework can be listed with their respective data types. We can figure out, for example, that the column "Supplies Subgroup" is an "object" data type and that the column "Client Size by Revenue" is an "integer data type." So, we have an understanding of columns that either contains integer values or string data.

Data Visualization

At this point, we are through with basic data exploration steps, so we will not attempt to build some appealing plots to portray the information visually and discover other concealed narratives from our data set.

Of all the available Python libraries providing data visualization features; "Seaborn" is one of the best available options, so we will be using the same. Make sure that the python plots module provided by "Seaborn" has been installed on your system and ready to be used. Now follow the steps below, generate the desired plot for the data set:

Step 1—Import the "Seaborn" module with the command "import seaborn as sns."

Step 2—Import the "Matplotlib" module with the command "import matplotlib.pyplot as plt."

Step 3—To set the "background color" of the plot as white, use command "sns.set(style="whitegrid", color_codes=True)."

Step 4—To set the "plot size" for all plots, use command "sns.set(rc={'figure.figsize':(11.7,8.27)})."

Step 5—To generate a "countplot", use command "sns.countplot('Route To Market', data=sales_data, hue = 'Opportunity Result')."

Step 6—To remove the top and bottom margins, use command "sns.despine(offset=10, trim=True)."

Step 7—To display the plot, use the command "plotplt.show()."

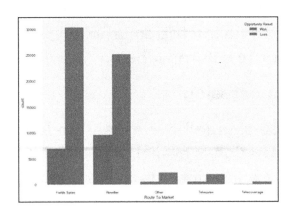

Quick recap—The "Seaborn" and "Matplotlib" modules were imported first. Then the "set()" technique was used to define the distinct characteristics for our plot, such as plot style and color. The background of the plot was defined to be white using the code snippet "sns.set(style= "whitegrid", color codes= True)." Then the plot size was define using command "sns.set(rc={'figure.figsize':(11.7,8.27)})" that define the size of the plot as "11.7px and 8.27px."

Next the command "sns.countplot('Route To Market',data= sales data, hue='Opportunity Result')" was used to generate the plot. The "countplot()" technique enables the creation of a count plot, which can expose multiple arguments to customize the count plot according to our requirements. As part of the first "countplot()" argument, the X-axis was defined as the column "Route to Market" from the data set. The next argument concerns the source of the data set, which would be "sales_data" data framework we imported earlier. The third argument is the color of the bar graphs that were defined as "blue" for the column labeled "won" and "green" for the column labeled "loss."

Data Pre-processing

By now, you should have a clear understanding of what information is available in the data set. From the data exploration step, we established that majority of the columns in our data set are "string data", but "Scikit-Learn" can only process numerical data. Fortunately, the Scikit-Learn library offers us many ways to convert string data into numerical data, for example, "LabelEncoder()" technique. To transform categorical labels from the data set such as "won" and "loss" into numerical values, we will use the "LabelEncoder()" technique.

Let's look at the pictures below to see what we are attempting to accomplish with the "LabelEncoder()" technique. The first image contains one column labeled "color" with three records namely, "Red", "Green" and "Blue." Using the "LabelEncoder()" technique, the record in the same "color" column can be converted to numerical values, as shown in the second image.

	Color
0	Red
1	Green
2	Blue

	Color
0	1
1	2
2	3

Let's begin the real process of conversion now. Using the "fit transform()" technique given by "LabelEncoder()," the labels in the categorical column like "Route To Market" can be encoded and converted to numerical labels comparable to those shown in the diagrams above. The function "fit transform()" requires input labels identified by the user and consequently returns encoded labels.

To know how the encoding is accomplished, let's go through an example quickly. The code instance below constitutes string data in the form of a list of cities such as ["Paris", "Paris", "Tokyo", "Amsterdam"] that will be encoded into something comparable to "[2, 2, 1,3]."

Step 1—To import the required module, use the command "from sklearn import preprocessing."

Step 2—To create the Label encoder object, use command "le = preprocessing.LabelEncoder()."

Step 3—To convert the categorical columns into numerical values, use command:

"encoded_value = le.fit_transform(["Paris", "Paris", "Tokyo", "Amsterdam"])"

"print(encoded_value) [1 1 2 0]"

And there you have it! We just converted our string data labels into numerical values. The first step was importing the preprocessing module that offers the "LabelEncoder()" technique. Followed by the development of an object representing the "LabelEncoder()" type. Then the "fit_transform()" function of the object was used to distinguish between distinct classes of the list ["Paris", "Paris", "Tokyo", "Amsterdam"] and output the encoded values of "[1 1 2 0]."

Did you observe that the "LabelEncoder()" technique assigned the numerical values to the classes in alphabetical order according to the initial letter of the classes, for example "(A)msterdam" was assigned code "0", "(P)aris" was assigned code "1" and "(T)okyo" was assigned code "2."

Creating Training and Test Subsets

To know the interactions between distinct characteristics and how these characteristics influence the target variable, an ML algorithm must be trained on a collection of information. We need to split the complete data set into two subsets to accomplish this. One subset will serve as the training data set, which will be used to train our algorithm to construct machine learning models. The other subset will serve as the test data set, which will be used to test the accuracy of the predictions generate by the machine learning model.

CHAPTER - 5

DATA STRUCTURES

Python is based on three reference structures: tuples, lists, and dictionaries. These structures are actually objecting that may contain other objects. They have quite different utilities and allow you to store information of all types.

These structures have a number of common features:

- To extract one or more objects from a structure, we always use the []

- For numerically indexed structures (tuples and lists), the structures are indexed to 0 (the first position is position 0)

The Tuples

This is a structure that groups multiple objects in indexed order. Its form is not modifiable (immutable) once created and is defined using parentheses. It has only one dimension. Any type of object can be stored in a tuple. For example, if

49

you want to create a tuple with different objects, we use:

tup1 = (1, True, 7.5.9)

You can also create a tuple by using the tuple () function. Access to the values of a tuple is done by the classical indexing of structures. Thus, if we want to access the third element of our tuple, we use:

In []: tup1 [2]

Out []: 7.5

Tuples can be interesting because they require little memory. Else, on the other hand, they are used as outputs of functions returning several values. Tuples as structures are objects. They have methods that are clean. These are few for a tuple:

In []: tup1.count (9)

Out []: 1

We often prefer lists that are more flexible.

Lists

The list is the reference structure in Python. It is modifiable and can contain any object.

Creating a list

We create a list using square brackets:

list1 = [3,5,6, True]

50

You can also use the list () function. The structure of a list is editable. It has many methods:

.append (): add value at the end of the list

.insert (i, val): insert value to the index i

.pop (i): retrieves the value of the index i

.reverse (): reverse the list

.extend (): extends the list with a list of values

Note—All of these methods modify the list, the equivalent in terms of classic code would be the following:

liste1.extend (list2)

equivalent to

list1 = list1 + list2

Lists have other methods including:

.index (val): returns the index of the value val

.count (val): returns the number of occurrences of val

.remove (val): remove the first occurrence of the value val from the list

Extract an Item from a List

As we have seen above, it is possible to extract an element using the brackets:

list1 [0]

We are often interested in the extraction of several elements. It is done by using the two points:

list1 [0: 2] or list1 [: 2]

In this example, we see that this system extracts two elements: the indexed element in 0 and the one indexed in position 1. So we have as a rule that i: j goes from the element I included in element j not included. Here are some other examples:

Extract the last element

list1 [-1]

Extract the last 3 elements list1 [-3: -1] or list1 [-3:]

A concrete example:

Suppose we wanted to create a list of countries. These countries are ordered in the list according to their population. We will try to extract the first three and the last three.

In []: country_list = ["China," "India," "United States," "France," "Spain," "Swiss"]

In []: print (country_list [: 3])

['China', 'India', 'United States']

In []: print (country_list [-3:])

['France', 'Spain', 'Switzerland']

In []: country_list.reverse ()

print (liste_pays)

['Switzerland', 'Spain', 'France', 'United States', 'India', 'China']

The Comprehension Lists

These are lists built iteratively. They are often very useful because they are more efficient than using loops to build lists. Here is a simple example:

In []: list_init = [4,6,7,8]

list_comp = [val ** 2 for val in list_init if val% 2 == 0]

The list comp_list allows you to store the even elements of list_init set to the square.

We will have:

In []: print (list_comp)

[16,36,64]

This notion of comprehension list is very effective. It avoids useless code (loops on a list) and performs better than creating a list iteratively. It also exists in dictionaries but not on tuples that are unchangeable. We will be able to use comprehension lists in the framework of the manipulation of data tables.

Strings—Character Lists

Strings in Python are encoded by default (since Python 3) in Unicode. You can declare a string of characters in three ways:

string1 = "Python for the data scientist"

string2 = 'Python for the data scientist'

string3 = """ "Python for the data scientist" """

The last one allows having strings on several lines. We will most often use the first. A string is actually a list of characters, and we will be able to work on the elements of a string as on those of a list:

In []: print (string1 [: 6])

print (string1 [-14:])

print (string1 [3:20 p.m.])

Python for the Data Scientist

Data strings can be easily transformed into lists:

In []: # we separate the elements using space

list1 = chaine1.split ()

print (list1)

['Python', 'for', 'the', 'Data', 'Scientist']

In []: # we join the elements with space

string1bis = "" .join (list1)

print (chaine1bis)

Dictionaries

The dictionaries constitute a third central structure to develop in Python. They allow key-value storage. So far, we have used items based on numerical indexing. So in a list, you access an element using its position list1 [0]. In a dictionary, we will access an element using a key defined when creating the dictionary. We define a dictionary with braces:

dict1 = {"cle1": value1, "cle2": value2, "cle3": value3}

This structure does not require any homogeneity of type in the values. From this, we can have a list like value1, a Boolean like value2, and an integer a value3.

To access an element of a dictionary, we use:

In []: dict1 ["cle2"]

Out []: value2

To display all the keys of a dictionary, we use:

In []: dict1.keys

Out []: ("cle1," "cle2," "cle3")

To display all the values of a dictionary, we use:

In []: dict1.items ()

Out []: (value1, value2, value3)

One can easily modify or add a key to a dictionary:

In []: dict1 ["key4"] = value4

You can also delete a key (and the associated value) in a dictionary:

In []: del dict1 ["cle4"]

As soon as you are more experienced in Python,

55

you will use more dictionaries. At first, we tend to favor lists dictionaries because they are often more intuitive (with numerical indexing). However, more expert Pythonist will quickly realize the usefulness of dictionaries. In particular, we will be able to store the data as well as the parameters of a model in a very simple way. Plus, the flexibility of Python's for loop adapts very well to dictionaries and makes them very effective when they are well built.

Programming

The Conditions

A condition in Python is very simple to implement; it is a keyword. As mentioned before, the Python language is based on the indentation of your code. We will use an offset for this indentation with four spaces. Fortunately, tools like Spyder or Jupyter notebooks will automatically generate this indentation.

Here is our first condition, which means: if a is true, then display "it is true":

if a is True:

print ("it's true")

There is no exit from the condition; it is the indentation that will allow us to manage it. Generally, we are also interested in the complement of this condition; we will use else for that:

if a is True:

print ("it's true")

else:

print ("it's not true")

We can have another case; if our variable a is not necessarily a Boolean, we use elif:

if a is True:

print ("it's true")

elif a is False:

print ("it's wrong")

else:

print ("it's not a Boolean")

The Loops

Loops are central elements of most programming languages. Python does not break this rule. However, you must be very careful with an interpreted language such as Python. Indeed, the treatment of loops is slow in Python, and we will use it in loops with few iterations. We avoid creating a loop repeating itself thousands of times on the lines of an array of data. However, we can use a loop on the columns of a data table to a few dozen columns.

The for loop

The Python loop has a somewhat specific format; it is a loop on the elements of a structure. We will write:

for elem in [1, 2]:

print (elem)

This piece of code will allow you to display 1 and 2. So the iterator of the loop (elem in our case) thus takes the values of the elements of the structure in the second position (after the in). These elements may be in different structures, but lists will generally be preferred.

Range, zip and enumerate functions

These three functions are very useful functions; they make it possible to create specific objects that may be useful in your code for your loops. The range () function is used to generate a sequence of numbers, starting from a given number or 0 by default and up to a number not included:

In []: print (list (range (5)))

[0, 1, 2, 3, 4]

In []: print (list (range (2,5)))

[2, 3, 4]

In []: print (list (range (2,15,2)))

[2, 4, 6, 8, 10, 12, 14]

We see here that the created range object can be easily transformed into a list with the list ().

In a loop, this gives:

for i in range (11):

print (i)

The zip and enumerate functions are also useful functions in loops and they use lists.

The enumerate () function returns the index and the element of a list. If we take our list of countries used earlier:

In []: for i, in enumerate (country_list):

print (i, a)

1. Swiss
2. Spain
3. France
4. United States
5. India
6. China

The zip function will allow linking many lists and simultaneously iterating elements of these lists.

If, for example, we want to simultaneously increment days and weather, we may use:

In []: for day, weather in zip (["Monday," "Tuesday"],

["beautiful," "bad"]):

print ("% s, it will make% s"% (day.capitalize (), weather))

Monday, it will be nice

Tuesday, it will be bad

In this code, we use zip () to take a pair of values at each iteration. The second part is a manipulation of the character strings. If one of the lists is longer than the other, the loop will stop as soon as it arrives at the end of one of them.

We can link enumerate and zip in one code, for example:

In []: for i, (day, weather) in enumerate (zip (["Monday," "Tuesday"], ["good," "bad"])):

print ("% i:% s, it will make% s"% (i, day.capitalize (), meteo))

0: Monday, it will be nice

1: Tuesday, it will be bad

We see here that i is the position of the element i.

Note—Replace in a string.

While loop

Python also allows you to use a while () loop that is less used and looks a lot like the while loop that we can cross in other languages. To exit this loop, we can use a station wagon with a condition. Warning, we must increment the index in the loop, at the risk of being in a case of an infinite loop.

We can have:

i = 1

while i <100:

i + = 1

if i> val_stop:

break

print (i)

This code adds one to i to each loop and stops when i reaches either val_stop, that is 100.

Note—The incrementation in Python can take several forms i = i + 1 or i + = 1.

Both approaches are equivalent in terms of performance; it's about choosing the one that suits you best.

```html
            <span class="stats-no" data-toggle="counter-up">232</span> Customers
        </div>
    </div>

    <div class="stats-col text-center col-md-3 col-sm-6">
        <div class="circle">
            <span class="stats-no" data-toggle="counter-up">79</span> Projects
        </div>
    </div>

    <div class="stats-col text-center col-md-3 col-sm-6">
        <div class="circle">
            <span class="stats-no" data-toggle="counter-up">1,463</span> Support
        </div>
    </div>

    <div class="stats-col text-center col-md-3 col-sm-6">
        <div class="circle">
            <span class="stats-no" data-toggle="counter-up">15</span> Hard Workers
        </div>
    </div>
</div>
</div>
</section>

<div class="block bg-primary block-pd-lg block-bg-overlay text-center" data-bg-i
data-toggle="parallax-bg">
    <h1>
        Welcome to a perfect
    </h1>
    <h>
```

CHAPTER - 6

DATA SCIENCE ALGORITHMS AND MODELS

The algorithms used in data science can be divided into several categories, mainly supervised learning, unsupervised learning, and to some degree, semi-supervised learning.

As the name suggests, supervised learning is aided by human interaction as the data scientist is required to provide the input and output in order to obtain a result from the predictions that are performed during the training process. Once the training is complete, the algorithm will use what it learned to apply to new but similar data.

We are going to focus on this type of learning algorithm. However, take note that their purpose is divided based on the problems they need to solve. Mainly there are two distinct categories, namely regression and classification. In the case of regression problems, your target is a numeric value, while in classification, it is a class or a label. To make things clearer, an example of a regression task is determining the average value of houses in a given city. A classification task, on the other hand, is supposed to take certain data like the petal and sepal length, and based on that information, determine what the species of a flower is.

With that in mind, let's start by discussing regression algorithms and how to work with them.

Regression

In data science, many tasks are resolved with the help of regression techniques. However, a regression can also be categorized into two different branches, which are linear regression and logistic regression. Each one of them is used to solve different problems; however, both of them are a perfect choice for prediction analyses because of the high accuracy of the results.

The purpose of linear regression is to shape a prediction value out of a set of unrelated variables. This means that if you need to discover the relationship between a number of variables, you can apply a linear regression algorithm to do the work for you. However, this isn't its main use. Linear regression algorithms are used for regression tasks. Keep in mind that logistic regression is not used to solve regression problems, as the name suggests. Instead, it is used for classification tasks.

With that being said, we are going to start by implementing a linear regression algorithm on the Boston housing dataset, which is freely available and even included in the Scikit-learn library. This dataset contains 506 samples, with 13 features and a numerical type target. We are going to break it into two sections, training and a testing set. There are no rules set in stone regarding the ratio of the split; however, it is generally accepted that it is best to keep the training set with a 70%-80% data distribution, and then save 20%-30% for the testing process.

K-Nearest Neighbors

This algorithm is one of the easiest ones to work with; however, it can solve some of the most challenging classification problems. The k-nearest neighbor algorithm can be used in various scenarios that require anything from compressing data to processing financial data. It is one of the most commonly used supervised machine learning algorithms, and you should do your best to practice your implementation technique.

The basic idea behind the algorithm is the fact that you should explore the relation between two different training observations. For instance, we will call them x and y, and if you have the input value of x, you can already predict the value of y. The way this works is by calculating the distance of a data point in relation to other data points. The k-nearest point is selected based on this distance, and then it is assigned to a specific class.

To demonstrate how to implement this algorithm, we are going to work with a much larger dataset than before; however, we will not use everything in it. Once again, we are going to rely on the Scikit-learn library in order to gain access to a dataset known as the MNIST handwritten digits dataset. This is, in fact, a database that holds roughly 70,000 images of handwritten digits, which are distributed in a training set with 60,000 images and a test set with 10,000 images. However, as already mentioned, we are not going to use the entire dataset because that would take too long for this demonstration. Instead, we will limit ourselves to 1000 samples. Let's get started:

In: from sklearn.utils import shuffle

from sklearn.datasets import

from sklearn.cross_validation import train_test_split

import pickle

mnist = pickle.load(open("mnist.pickle", "rb"))

mnist.data, mnist.target = shuffle(mnist.data, mnist.target)

As usual, we first import the dataset and the tools we need. However, you will notice one additional step here, namely object serialization. This means we converted an object to a different format so that it can be used later, but also reverted back to its original version if needed. This process is referred to as pickling, and that is why we have the seemingly out of place pickle module imported. This will allow us to communicate objects through a network if needed. Now, let's cut through the dataset until we have only 1000 samples:

mnist.data = mnist.data[:1000]

mnist.target = mnist.target[:1000]

X_train, X_test, y_train, y_test = train_test_split(mnist.data,

mnist.target, test_size=0.8, random_state=0)

In: from sklearn.neighbors import KNeighborsClassifier

KNN: K=10, default measure of Euclidean distance

clf = KNeighborsClassifier(3)

clf.fit(X_train, y_train)

y_pred = clf.predict(X_test)

Now let's see the report with the accuracy metrics like earlier:

In: from sklearn.metrics import classification_report

print (classification_report(y_test, y_pred))

And here are the results:

Out:

PRECISION	RECALL	F1-SCORE	SUPPORT	
0.0	0.68	0.90	0.78	79
1.0	0.66	1.00	0.79	95
2.0	0.83	0.50	0.62	76
3.0	0.59	0.64	0.61	85
4.0	0.65	0.56	0.60	75
5.0	0.76	0.55	0.64	80
6.0	0.89	0.69	0.77	70
7.0	0.76	0.83	0.79	76
8.0	0.91	0.56	0.69	77
9.0	0.61	0.75	0.67	87
AVG / TOTAL	0.73	0.70	0.70	800

The results aren't the best; however, we have only implemented the "raw" algorithm without performing any kind of preparation operations that would clean and denoise the data. Fortunately, the training speed was excellent, even at this basic level. Remember, when working with supervised algorithms or any algorithms for that matter, you are always trading accuracy for processing speed or vice versa.

Support Vector Machines

The SVM is one of the most popular supervised learning algorithms due to its capability of solving both regressions as well as classification problems. In addition, it has the ability to identify outliers as well. This is one all-inclusive data science algorithm that you cannot miss. So what's so special about this algorithm?

First of all, support vector machines don't need much processing power in order to keep up with the prediction accuracy. This algorithm, however, is in a league of its own, and you won't have to worry too much about sacrificing training speed for the accuracy or the other way around. Furthermore, support vector machines can be used to eliminate some of the noise as well while performing the regression or classification tasks.

This type of algorithm has many real-world applications, and that is why it is important for you to understand its implementation. It is used in facial recognition software, text classification, handwriting recognition software, and so on. The basic concept behind it, however, simply involves the distance between the nearest points where a hyperplane is selected from the margin between a number of support vectors. Take note that what is known as a hyperplane here is the object that divides the information space for the purpose of classification.

To put all of this theory in the application, we are going to rely on the Scikit-learn library once again. The algorithm will be implemented in such a way to demonstrate the accuracy of the prediction in the case of identifying real banknotes. We mentioned earlier that support vector machines are effective when it comes to image classification. Therefore, this algorithm is perfectly suited for our goals. What we need to solve in this example is a simple binary classification problem because we need to train the algorithm to determine whether the banknote is valid or not.

The bill will be described using several attributes. Keep in mind that unlike the other classification algorithms, a support vector machine determines its decision limit by defining the maximum distance between the data points which are nearest to the relevant classes. However, we aren't looking to limit the decision, we just want to find the best one. The nearest points in this best decision are what we refer to as support vectors. With that being said, let's import a new dataset and several tools:

```python
import numpy as np
import pandas as pd
import matplotlib.pyplot as plt
dataset = pd.read_csv ("bank_note.csv")
```

As usual, the first step is learning more about the data we are working with. Let's learn how many rows and columns we have and then obtain the data from the first five rows only:

```python
print (dataset.shape)
print (dataset.head())
```

Here's the result:

	VARIANCE	SKEWNESS	CURTOSIS	ENTROPY	CLASS
0	3.62160	8.6661	-2.8073	-0.44699	0
1	4.454590	8.1674	-2.4586	-1.46210	0
2	3.86600	-2.6383	1.9242	0.10645	0
3	3.45660	9.5228	-4.0112	-3.59440	0
4	0.32924	-4.4552	4.5718	-0.98880	0

Now we need to process this information in order to establish the training and testing sets. This means that we need to reduce the data to attributes and labels only:

x = dataset.drop ('Class', axis = 1)

y = dataset ['Class']

The purpose of this code is to store the column data as the x variable and then apply the drop function in order to avoid the class column so that we can store it inside a 'y' variable. By reducing the dataset to a collection of attributes and labels, we can start defining the training and testing data sets. Split the data just like we did in all the earlier examples. Next, let's start implementing the algorithm.

We need Scikit-learn for this step because it contains the support vector machine algorithm, and therefore we can easily access it without requiring outside sources.

from sklearn.svm import SVC

svc_classifier = SVC (kernel = 'linear')

svc_classifier.fit (x_train, y_train)

pred_y = svc.classifier.predict(x_test)

Finally, we need to check the accuracy of our implementation. For this step, we are going to use a confusion matrix, which will act as a table that displays the accuracy values of the classification's performance. You will see a number of true positives, true negatives, as well as false positives and false negatives. The accuracy value is then determined from these values. With that being said, let's take a look at the confusion matrix and then print the classification report:

from sklearn.metric import confusion_matrix

print (confusion_matrix (y_test, pred_y)

This is the output:

[[160 1]

 [1   113]]

Accuracy Score: 0.99

Now let's see the familiar classification report:

from sklearn.metrics import classification_report

print (classification_report(y_test, y_pred))

And here are the results of the report:

PRECISION	RECALL	F1-SCORE	SUPPORT	
0.0	0.99	0.99	0.99	161
1.0	0.99	0.99	0.99	114
AVG / TOTAL	0.99	0.99	0.99	275

73

Based on all of these metrics, we can determine that we obtained a very high accuracy with our implementation of the support vector machines. A score of 0.99 is almost as good as it can get; however, there is always room for improvement.

CHAPTER - 7
DATA AGGREGATION AND GROUP OPERATIONS

This represents the first part of aggregation and clustering using Pharo DataFrame. This will only handle the basic functionality like clustering a data series using values of a separate series of the corresponding size and using aggregation functions to the grouped data structures.

The next iterations will deal with functionality extended based on the targeted scenarios. The implementation is likely to change into something optimized.

Definition of Data Frame

This represents spreadsheets such as data structures that deliver an API for cleaning, slicing, and analyzing data.

In case you want to read more about the DataFrame project, you need to consider the documentation.

Split-Apply-Combine

The split-apply-combine is a technique where you categorize a certain task into manageable parts and then integrate all the parts together.

The data aggregation and grouping facilitates the production of summaries for analysis and display. For example, when you calculate the average values or creating a table of counts. This is a step that adheres to the split-apply-combine procedure.

1. Separate the data into sections based on a given procedure.
2. Use the function to every cluster independently.
3. Combine the results using a data structure.

Implementation

In this part, you will discover how the grouping and aggregation function is being implemented. In case you don't want these details, you can skip to the next part.

Take, for instance, this message that has been sent to firstSeries object:

```
firstSeries groupBy: secondSeries.
```

Once this message is sent, firstSeries will define an object of DataSeriesGrouped, which divides firstSeries into various subseries depending on the values of secondSeries.

The collection of subseries is later kept as an object of DataSeries whose keys are equivalent to the special values of the secondSeries and values store the subseries of firstSeries. That will match each of those unique values.

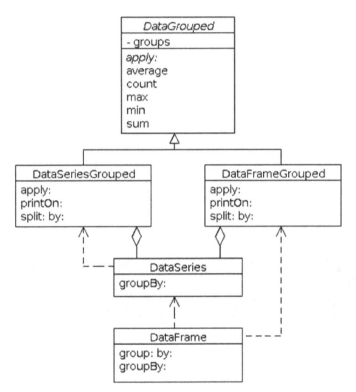

This means that the groups represent DataSeries that contain keys that match the unique values contained in a string in which the data frame is recognized. When the data frame is categorized by a single column, that column is removed from the data frame before grouping. Therefore, this eliminates data duplication because the same values will be preserved as keys.

In the case of DataSeriesGrouped, each subsystem will connect to a scalar, and all subsequent scalars will be merged into a DataSeries. When it comes to DataFrameGrouped, it will include the block in each column of each subdirectory box and display the final scalar table as the new DataFrame.

The combination is done with the use of messages. It requires a block as an argument and uses it on every value in the group string and then integrates it into a new data structure.

The most common aggregation functions, such as average, minimum, and maximum, deliver smaller messages. In the next iteration, these messages are useful and act as shortcuts.

```
average
    self apply: [ :each   each average ].
```

However, these messages will carry the optimized implementations of the likened aggregations because it is necessary that these functions are time and memory efficient.

Let's examine the grouping series.

The easiest example of using this groupBy operator is to classify the values of a series using values of the same size.

```
bill : tips column:
sex  : tips column:
```

The result of the above query will be an object. This object will separate the bill into two series.

Because a lot of time, you need to classify the group series that resemble columns of a single data frame. There is a useful shortcut.

How to Group Data Frames?

Besides the shortcut for classifying columns. The DataFrame has a method for classifying one of its columns.

The response of the above query will be an object of DataFrameGrouped, keeping two different data frames for smokers and non-smokers.

The smoker column will be removed from the above data frames because its values will be kept as keys within a DataFrameGrouped object. Additionally, the different groups of smokers and non-smokers will enable the complete reconstruction of the smoker column when needed.

The aggregation functions represent the ones that accept different input and display a scalar value that sums up the values of that particular series. These refer to statistical functions such as min, max, stdev, and many more.

Once the data has been combined, next, you can

use the aggregation function to get the integrated data structure that sums up the original data.

```
grouped := tips group: #total_bill by: #day.
grouped apply: [ :each | each average round: 2].
```

Since the grouping is being done to a column of DataFrame by a separate column. The result will be a DataSeries object.

As said before, the DataGrouped presents shortcuts for popularly applied aggregation functions such as count, sum, min, and max. At the moment, these are shortcuts, but in the future, they will execute the optimized aggregations that will be used faster.

Once the data box has been grouped into a DataFrameGrouped object, we can also apply an aggregate function to that object. DataFrameGrouped implements the apply message: for the function to apply to every column in every child data frame, producing the incremental value. These steps are then combined into a new data frame.

The result of this query will be a data box containing the number of empty cells for each column, corresponding to 'Male' and 'Female' rows.

```
        | total_bill   tip   smoker   day   time   size
--------+---------------------------------------------------
Female  | 87           87    87       87    87     87
Male    | 157          157   157      157   157    157
```

80

CHAPTER - 8

PRACTICAL CODES AND EXERCISES TO USE PYTHON

We will do a few different Python exercises here so that you can have a little bit of fun and get a better idea of how you would use the different topics that we have talked about in this guidebook to your benefit. There are a lot of neat programs that you can use when you write in Python, but the one here will give you a good idea of how to write codes and how to use the examples that we talked about in this guidebook in real coding.

Creating a Magic 8 Ball

The first project that we are going to take a look at here is how to create your own Magic 8 ball. This will work just like a regular magic 8 ball, but it will be on the computer. You can choose how many answers that you would like to have available to those who are using the program, but we are going to focus on having eight responses show up for the user at a random order, so they get something different each time.

Setting up this code is easier than you think. Take some time to study this code, and then write it out into the compiler. The code that you need to use to create a program that includes your own Magic 8 ball will include:

```python
# Import the modules

import sys

import random

ans = True

while ans:

    question = raw_input("Ask the magic 8 ball a question: (press enter to quit)")

    answers = random.randint(1,8)

    if question == "":

        sys.exit()

    elif answers ==1:

        print("It is certain")

    elif answers == 2:

        print("Outlook good")

    elif answers == 3:

        print("You may rely on it")

    elif answers == 4:
```

print("Ask again later")

elif answers == 5:

print("Concentrate and ask again")

elif answers == 6:

print("Reply hazy, try again.")

elif answers == 7:

print("My reply is no")

elif answers == 8:

print("My sources say no")

Remember, in this program, we chose to go with eight options because it is a Magic 8 ball, and that makes the most sense. But if you would like to add in some more options, or work on another program that is similar and has more options, then you would just need to keep adding in more of the elif statement to get it done. This is still a good example of how to use the elif statement that we talked about earlier and can give us some good practice on how to use it. You can also experiment a bit with the program to see how well it works and make any changes that you think are necessary to help you get the best results.

How to Make a Hangman Game

The next project that we are going to take a look at is creating your own Hangman game. This is a great game to create because it has a lot of the different options that we have talked about throughout this guidebook and can be a great way to get some practice on the various topics that we have looked at. We are going to see things like a loop present, some comments, and more, and this is a good way to work with some of the conditional statements that show up as well.

Now, you may be looking at this topic and thinking it is going to be hard working with a Hangman game. It is going to have a lot of parts that go together as the person makes a guess and the program tries to figure out what is going on, whether the guesses are right, and how many chances the user gets to make these guesses. But using a lot of the different parts that we have already talked about in this guidebook can help us to write out this code without any problems. The code that you need to use to create your very own Hangman game in Python includes:

importing the time module

importing time

#welcoming the user

Name = raw_input("What is your name?")

print("Hello, + name, "Time to play hangman!")

print("

```
#wait for 1 second
time.sleep(1)
print("Start guessing...")
time.sleep(.05)
#here we set the secret
word = "secret"
#creates a variable with an empty value
guesses = ' '
#determine the number of turns
turns = 10
#create a while loop
#check if the turns are more than zero
while turns > 0:
    #make a counter that starts with zero
    failed = 0
    #for every character in secret_word
    for car in word:
        #see if the character is in the players guess
        if char in guesses:
            #print then out the character
```

```
        print char,
    else
        # if not found, print a dash
        print "_",
        # and increase the failed counter with one
        failed += 1
    #if failed is equal to zero
    #print You Won
    if failed == 0:
        print("You Won")
        #exit the script
        Break
    print
    # ask the user to guess a character
    guess = raw_input("guess a character:")
    #set the players guess to guesses
    guesses += guess
    # if the guess is not found in the secret word
    if guess not in word:
        #turns counter decreases with 1 (now 9)
        turns -= 1
```

86

```
#print wrong

print("Wrong")

# how many turns are left

Print("You have," + turns, 'more guesses')

#if the turns are equal to zero

if turns == 0

#print "You Lose"
```

This is a longer piece of code, especially when it is compared to the Magic 8 Ball that we did above, but take a deep breath, and go through it all to see what you recognize is there. This isn't as bad as it looks, and much of it is comments to help us see what is going on at some of the different parts of the code. This makes it easier to use for our own needs and can ensure that we know what is going on in the different parts.

Making Your Own K-Means Algorithm

Now that we have had some time to look at a few fun games and examples that you can do with the help of the Python code, let's take a moment to look at some of the things that you can do with Machine Learning and artificial intelligence with your coding. We spent some time talking about how you can work with these and some of the different parts of the code, as well as how Python is going to work with the idea of Machine Learning.

And now, we are going to take that information and create one of our Machine Learning algorithms to work with as well.

Before we work on a code for this one, we need to take a look at what this k-means clustering means. This is a basic algorithm that works well with Machine Learning and is going to help you to gather up all of the data that you have in your system, the data that isn't labeled at the time, and then puts them all together in their little group of a cluster.

The idea of working with this kind of cluster is that the objects that fall within the same cluster, whether there are just two or more, are going to be related to each other in some manner or another, and they are not going to be that similar to the data points that fall into the other clusters. The similarity here is going to be the metric that you will want to use to show us the strength that is in the relationship between the two.

When you work on this particular algorithm, it is going to be able to form some of the clusters that you need of the data, based on how similar the values of data that you have. You will need to go through and give them a specific value for K, which will be how many clusters that you would like to use. It is best to have at least two, but the number of these clusters that you work with will depend on how much data you have and how many will fit in with the type of data that you are working with.

With this information in mind and a good background of what the K-means algorithm is going to be used for, it is time to explore a bit more about how to write your own codes and do an example that works with K-means. This helps us to practice a bit with Machine Learning and gives us a chance to practice some of our own new Python skills.

```
import numpy as np

import matplotlib.pyplot as plt

def d(u, v):

    diff = u - v

    return diff.dot(diff)

def cost(X, R, M):

    cost = 0

    for k in xrange(len(M)):

        for n in xrange(len(X)):

            cost += R[n,k]*d(M[k], X[n])

    return cost
```

After this part, we are going to take the time to define your function so that it is able to run the k-means algorithm before plotting the result. This is going to end up with a scatterplot where the color will represent how much of the membership is inside of a particular cluster. We would do that with the following code:

```python
def plot_k_means(X, K, max_iter=20, beta=1.0):
    N, D = X.shape
    M = np.zeros((K, D))
    R = np.ones((N, K)) / K

    # initialize M to random
    for k in xrange(K):
        M[k] = X[np.random.choice(N)]

    grid_width = 5
    grid_height = max_iter / grid_width
    random_colors = np.random.random((K, 3))
    plt.figure()

    costs = np.zeros(max_iter)
    for i in xrange(max_iter):
        # moved the plot inside the for loop
        colors = R.dot(random_colors)
        plt.subplot(grid_width, grid_height, i+1)
        plt.scatter(X[:,0], X[:,1], c=colors)

        # step 1: determine assignments / responsibilities
```

90

```python
# is this inefficient?
for k in xrange(K):
    for n in xrange(N):
        R[n,k] = np.exp(-beta*d(M[k], X[n])) / np.sum ( np.exp(-beta*d(M[j], X[n])) for j in xrange(K) )

# step 2: recalculate means
for k in xrange(K):
    M[k] = R[:,k].dot(X) / R[:,k].sum()

costs[i] = cost(X, R, M)
f i > 0:
    if np.abs(costs[i] - costs[i-1]) < 10e-5:
        break
plt.show()
```

Notice here that both the M and the R are going to be matrices. The R is going to become the matrix because it holds onto 2 indices, the k and the n. M is also a matrix because it is going to contain the K individual D-dimensional vectors. The beta variable is going to control how fuzzy or spread out the cluster memberships are and will be known as the hyperparameter. From here, we are going to create a main function that will create random clusters and then call up the functions that we have already defined above.

```python
def main():
    # assume 3 means
    D = 2 # so we can visualize it more easily
    s = 4 # separation so we can control how far apart the means are
    mu1 = np.array([0, 0])
    mu2 = np.array([s, s])
    mu3 = np.array([0, s])

    N = 900 # number of samples
    X = np.zeros((N, D))
    X[:300, :] = np.random.randn(300, D) + mu1
    X[300:600, :] = np.random.randn(300, D) + mu2
    X[600:, :] = np.random.randn(300, D) + mu3

    # what does it look like without clustering?
    plt.scatter(X[:,0], X[:,1])
    plt.show()

    K = 3 # luckily, we already know this
    plot_k_means(X, K)

    # K = 5 # what happens if we choose a "bad" K?
    # plot_k_means(X, K, max_iter=30)

    # K = 5 # what happens if we change beta?
```

```
    # plot_k_means(X, K, max_iter=30, beta=0.3)

if __name__ == '__main__':

    main()
```

Yes, this process is going to take some time to write out here, and it is not always an easy process when it comes to working through the different parts that come with Machine Learning and how it can affect your code. But when you are done, you will be able to import some of the data that your company has been collecting, and then determine how this compares using the K-means algorithm as well.

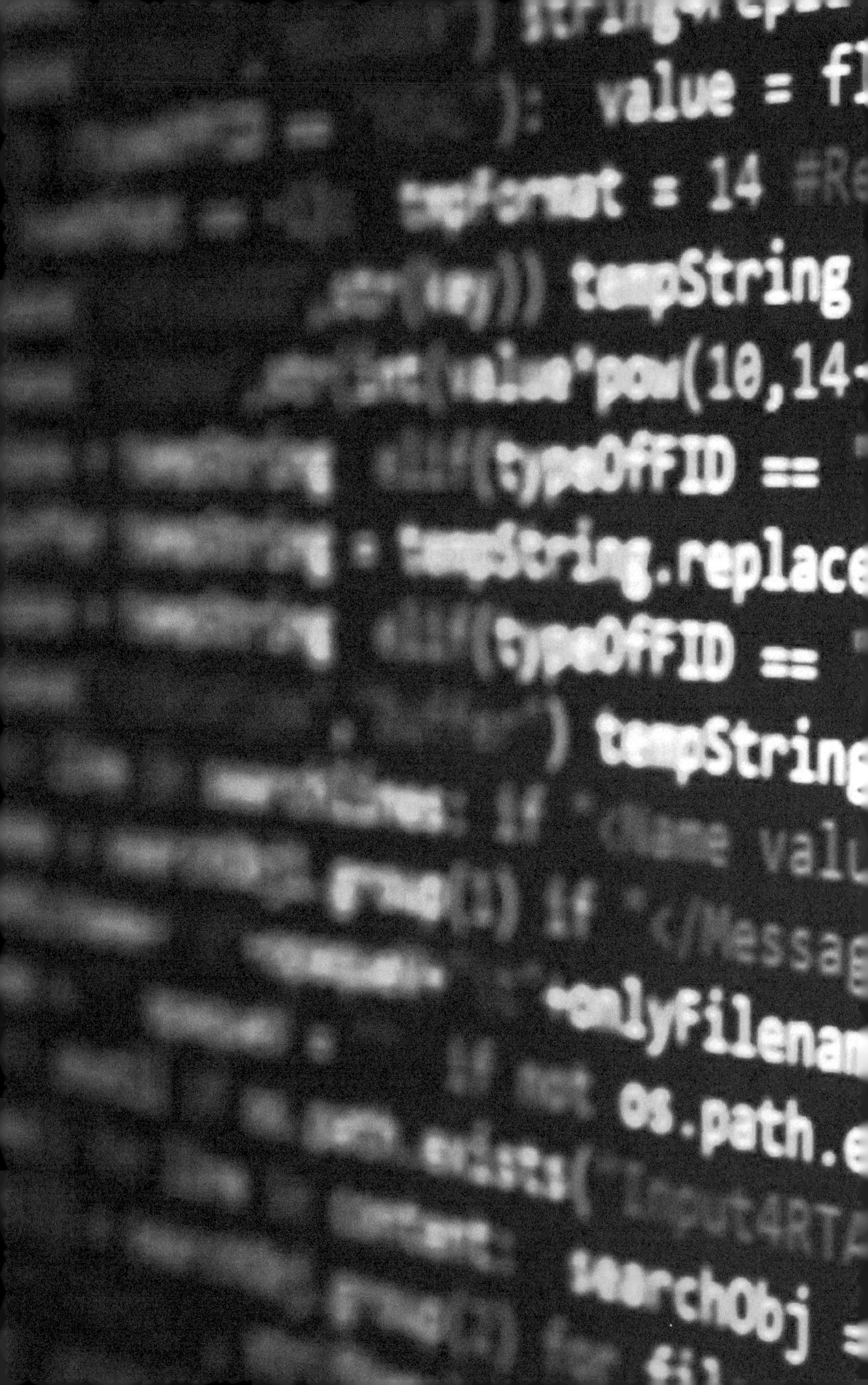

CHAPTER - 9

FUNCTIONS AND MODULES IN PYTHON

When you are working with a language like Python, there will be times when you will need to work with something that is known as a function. These functions are going to be blocks of reusable code that you will use in order to get your specific tasks done. But when you define one of these functions in Python, we need to have a good idea of the two main types of functions that can be used, and how each of them work. The two types of functions that are available here are known as built-in and user-defined.

The built-in functions are the ones that will come automatically with some of the packages and libraries that are available in Python. However, we are going to spend our time working with the user-defined functions because these are the ones that the developer will create and use for special codes they write. In Python though, one thing to remember, no matter what kind of function you are working with, is that all of them will be treated like objects. This is good news because it can make it a lot easier to work with these functions compared to other coding languages.

Built-in Functions				
abs()	divmod()	input()	open()	staticmethod()
all()	enumerate()	int()	ord()	str()
any()	eval()	isinstance()	pow()	sum()
basestring()	execfile()	issubclass()	print()	super()
bin()	file()	iter()	property()	tuple()
bool()	filter()	len()	range()	type()
bytearray()	float()	list()	raw_input()	unichr()
callable()	format()	locals()	reduce()	unicode()
chr()	frozenset()	long()	reload()	vars()
classmethod()	getattr()	map()	repr()	xrange()
cmp()	globals()	max()	reversed()	zip()
compile()	hasattr()	memoryview()	round()	__import__()
complex()	hash()	min()	set()	
delattr()	help()	next()	setattr()	
dict()	hex()	object()	slice()	
dir()	id()	oct()	sorted()	

The user-defined functions that we are going to talk about in the next section are going to be important and can really expand out some of the work that we are doing as well. We also need to take a look at some of the work that we are able to do with built-in functions as well. The list above includes many of the ones that are found in the Python language. Take some time to study them and see what they are able to do to help them get things done.

Why are User-defined Functions So Important?

To keep it simple, a developer is going to have the option of either writing out some of their own functions, known as a user-defined function, or they are able to go through and borrow a function from another library, one that may not be directly associated with Python. These functions are sometimes going to provide us with a few advantages, depending on how and when we would like to use them in the code. Some of the things that we need to remember when working on these user-defined functions, to gain a better understanding of how they work, will include:

These functions are going to be made out of code blocks that are reusable. It is necessary to only write them out once. Then you can use them as many times as you need in the code. You can even make use of user-defined functions in some of your other applications as well.

These functions can also be very useful. You can use them to help with anything that you want—from writing out specific logic in business to working on common utilities. You can also modify them based on your own requirements, to make the program work properly.

The code is often going to be friendly for developers, easy to maintain, and well-organized. This means that you are able to support the approach for modular design.

You are able to write out these types of functions independently. Also, the tasks of your project can be distributed for rapid application development, if needed.

A user-defined function that is thoughtfully well-defined can help ease the process for the development of an application.

Now that we know a little bit more about the basics of a user-defined function, it is time to look at some of the different arguments that can come with these functions before moving on to some of the codes that you can use with this kind of function.

Options for Function Arguments

Any time that you are ready to work with these kinds of functions in your code, you will find that they have the ability to work with four types of arguments. These arguments and the meanings behind them will be pre-defined, and the developer is not always going to be able to change them up. Instead, the developer is going to have the option to use them, but also follow the rules that are there with them. You do get the option to add a bit to the rules to make the functions work the way that you want. As we said before, there are four argument types you can work with, and these include:

- **Default Arguments:** In Python, we are going to find that there is a different way to represent the default values and the syntax for the arguments of your functions. These default values are going to indicate that the argument of the function is going to take that value if you do not have a value for the argument which can pass through the call of the function. The best way to figure out where the default value will be is to look for the equal sign.

- **Required Argument:** These are the kinds of arguments that will be mandatory to the function that you are working on. These values need to go through and be passed in the right order and number, either when the function is called out or when the code will not be able to run the right way.

- **Keyword Arguments:** These are going to be the argument that will be able to help with the function call inside Python. These keywords are going to be the ones that we mention through the function call, along with some of the values that will go through this one. These keywords will be mapped with the function argument so that you are able to identify all of the values, even if you do not keep the order the same when the code is called.

- **Variable Arguments:** The last argument that we are going to take a look at here is the variable number of arguments. This is a good one to work with when you are not sure how many arguments are going to be necessary for the code that you are writing to pass the function. Otherwise, you can use this to design your code where any number of arguments can be passed, as long as they have been able to pass any requirements in the code that you set.

Writing a Function

Now that we have a little better idea of what these functions are like, and some of the argument types that are available in Python, it is time for us to learn the steps that you need to accomplish all of this. There are going to be four basic steps that we are able to use to make all of this happen, and it is really up to the programmer of how difficult or simple you would like this to be. We will start out with some of the basics, and then you can go through and make some adjustments as needed. Some of the steps that we need to take in order to write out our own user-defined functions are given below.

- Declare your function. You will need to use the 'def' keyword, and then have the name of the function come right after it.

- Write out the arguments. These need to be inside the two parentheses of the function. End this declaration with a colon, to keep up with the proper writing protocol in this language.

- Add in the statements that the program is supposed to execute at this time.

- End the function. You can choose whether you would like to do it with a return statement.

An example of the syntax that you would use when you want to make one of your own user-defined functions is:

```
def userDefFunction (arg1, arg2, arg3, ...):

    program statement1

    program statement2

    program statement3

    ....

    Return;
```

Working with functions can be a great way to ensure that your code is going to behave the way that you would like it to. Making sure that you get it set up in the proper manner can be really important as well. There are many times when the functions will come out and serve some purpose, so taking the time now to learn how to use them can be very important to the success of your code.

Python Modules

Modules consist of definitions as well as program statements.

An illustration is a file name config.py, which is considered as a module. The module name would be config. Modules are used to help break large programs into smaller, manageable, and organized files, as well as promoting reusability of code.

Example:

Creating the First module

Start IDLE.

Navigate to the File menu and click New Window.

Type the following:

Def add(x, y):(This is a program to add two numbers and return the outcome)

Outcome: x+y

Return outcome

Import Module

The keyword "import" is used to import.

Example:

Import first

The dot operator can help us access a function as long as we know the name of the module.

Example:

Start IDLE.

Navigate to the File menu and click New Window.

Type the following:

first.add(6,8)

Import statement in Python

The import statement can be used to access the definitions within a module via the dot operator.

Start IDLE.

Navigate to the File menu and click New Window.

Type the following:

import math

print("The PI value is", math.pi)

Import with renaming

Example:

Start IDLE.

Navigate to the File menu and click New Window.

Type the following:

import math as h

print("The PI value is-",h.pi)

In this case, h is our renamed math module that saves typing time in some instances. When we rename, the new name becomes the valid and recognized one instead of the original one.

From... import statement Python

It is possible to import particular names from a module, rather than importing the entire module.

Example:

Start IDLE.

Navigate to the File menu and click New Window.

Type the following:

from math import pi

print("The PI value is-", pi)

Importing all names

Example:

Start IDLE.

Navigate to the File menu and click New Window.

Type the following:

from math import*

print("The PI value is-", pi)

In this context, we are importing all definitions from a particular module. However, it is an encouraged norm as it can lead to unseen duplicates.

Module Search Path in Python

Example:

Start IDLE.

Navigate to the File menu and click New Window.

Type the following:

import sys

sys.path

Python searches everywhere, including the .sys file.

Reloading a Module

Python will only import a module once, increasing efficiency in execution.

print("This program was executed") import mine

Reloading Code

Example:

Start IDLE.

Navigate to the File menu and click New Window.

Type the following:

import mine

import mine

import mine

mine.reload(mine)

Dir() built-in Python function

For discovering names contained in a module, we use the dir() inbuilt function.

Syntax

dir(module_name)

Python Package

Files in Python hold modules, and directories are stored in packages. A single package in Python holds similar modules. Therefore, different modules should be placed in different Python packages.

CHAPTER - 10
DATA SCIENCE AND THE CLOUD

Data science is a mixture of many concepts. To become a data scientist, it is important to have some programming skills. Even though you might not know all the programming concepts related to infrastructure, but having basic skills in computer science concepts is a must. You must install the two most common and most used programming languages, i.e., R and Python, on your computer. With the ever-expanding advanced analytics, Data Science continues to spread its wings in different directions. This requires collaborative solutions like predictive analysis and recommendation systems. Collaboration solutions include research and notebook tools integrated with code source control. Data science is also related to the cloud. The information is also stored in the cloud. So, this lesson will enlighten you with some facts about the "data in the Cloud." So let us understand what cloud means and how the data is stored and how it works.

What Is the Cloud?

The cloud can be described as a global server network, each having different unique functions. Understanding networks is required to study the cloud. Networks can be simple or complex clusters of information or data.

Network

As specified earlier, networks can have a simple or small group of computers connected or large groups of computers connected. The largest network can be the Internet. The small groups can be home local networks like Wi-Fi and Local Area Network that is limited to certain computers or locality. There are shared networks such as media, web pages, app servers, data storage, and printers, and scanners. Networks have nodes, where a computer is referred to as a node. The communication between these computers is established by using protocols. Protocols are the intermediary rules set for a computer. Protocols like HTTP, TCP, and IP are used on a large scale. All the information is stored on the computer, but it becomes difficult to search for information on the computer every time. Such information is usually stored in a data Centre. Data Centre is designed in such a way that it is equipped with support security and protection for the data. Since the cost of computers and storage has decreased substantially, multiple organizations opt to make use of multiple computers that work together that one wants to scale. This differs from other scaling solutions like buying other computing devices. The intent behind this is to keep the work going continuously even if a computer fails; the other will continue the operation. There is a need to scale some cloud applications, as well. Having a broad look at some computing applications like YouTube, Netflix, and Facebook that requires some scaling. We rarely experience such applications failing, as they have set up their systems on the cloud. There is a network cluster in the cloud, where many computers are connected to the same networks and accomplish similar tasks. You can call it as a single source of information or a single computer that manages everything to improve performance, scalability, and availability.

Data Science in the Cloud

The whole process of Data Science takes place in the local machine, i.e., a computer or laptop provided to the data scientist. The computer or laptop has inbuilt programming languages and a few more prerequisites installed. This can include common programming languages and some algorithms. The data scientist later has to install relevant software and development packages as per his/her project. Development packages can be installed using managers such as Anaconda or similar managers. You can opt for installing them manually too. Once you install and enter into the development environment, then your first step, i.e., the workflow, starts where your companion is only data. It is not mandatory to carry out the task related to Data Science or Big data on different development machines. Check out the reasons behind this:

1. The processing time required to carry out tasks on the development environment fails due to processing power failure.

2. Presence of large data sets that cannot be contained in the development environment's system memory.

3. Deliverables must be arrayed into a production environment and incorporated as a component in a large application.

4. It is advised to use a machine that is fast and powerful.

Data scientist explores many options when they face such issues; they make use of on-premise machines or virtual machines that run on the cloud. Using virtual machines and auto-scaling clusters has various benefits, such as they can span up and discard it anytime in case it is required. Virtual machines are customized in a way that will fulfill one's computing power and storage needs. Deployment of the information in a production environment to push it in a large data pipeline may have certain challenges. These challenges are to be understood and analyzed by the data scientist. This can be understood by having a gist of software architectures and quality attributes.

Software Architecture and Quality Attributes

A cloud-based software system is developed by Software Architects. Such systems may be a product or service that depends on the computing system. If you are building software, the main task includes the selection of the right programming language that is to be programmed. The purpose of the system can be questioned; hence, it needs to be considered. Developing and working with software architecture must be done by a highly skilled person. Most of the organizations have started implementing effective and reliable cloud environment using cloud computing. These cloud environments are deployed over to various servers, storage, and networking resources. This is used in abundance due to its less cost and high ROI.

The main benefit to data scientists or their teams is that they are using the big space in the cloud to explore more data and create important use cases. You can release a feature and have it tested the next second and check whether it adds value or it is not useful to carry forward. All this immediate action is possible due to cloud computing.

Sharing Big Data in the Cloud

The role of Big Data is also vital while dealing with the cloud as it makes it easier to track and analyze insights. Once this is established, big data creates great value for users.

The traditional way was to process wired data. It became difficult for the team to share their information with this technique. The usual problems included transferring large amounts of data and collaboration of the same. This is where cloud computing started sowing its seed in the competitive world. All these problems were eliminated due to cloud computing, and gradually, teams were able to work together from different locations and overseas as well. Therefore, cloud computing is very vital in both Data Science as well as Big data. Most of the organizations make use of the cloud. To illustrate, a few companies that use the cloud are Swiggy, Uber, Airbnb, etc. They use cloud computing for sharing information and data.

Cloud and Big Data Governance

Working with the cloud is a great experience as it reduces resource cost, time, and manual efforts. But the question arises that how organizations deal with security, compliance, governance? Regulation of the same is a challenge for most companies. Not limited to Big data problems, but working with the cloud also has its issues related to privacy and security. Hence, it is required to develop a strong governance policy in your cloud solutions. To ensure that your cloud solutions are reliable, robust, and governable, you must keep it as an open architecture.

Need for Data Cloud Tools to Deliver High Value of Data

The demand for a data scientist in this era is increasing rapidly. They are responsible for helping big and small organizations to develop useful information from the data or data set that is provided. Large organizations carry massive data that need to analyze continuously. As per recent reports, almost 80% of the unstructured data received by the organizations are in the form of social media, emails, i.e., Outlook, Gmail, etc., videos, images, etc. With the rapid growth of cloud computing, data scientists deal with various new workloads that come from IoT, AI, Blockchain, Analytics, etc. Pipeline. Working with all these new workloads requires a stable, efficient, and centralized platform across all teams. With all this, there is a need for managing and recording new data as well as legacy documents. Once a data scientist is given a task, and he/she has the dataset to work on, he/she must possess the right skills to analyze the ever-increasing volumes through cloud technologies. They need to convert the data into useful insights that would be responsible for uplifting the business. The data scientist has to build an algorithm and code the program. They mostly utilize 80% of their time to gathering information, creating and modifying data, cleaning if required, and organizing data. Rest 20% is utilized for analyzing the data with effective programming. This calls for the requirement of having specific cloud tools to help the data scientist to reduce their time searching for appropriate information. Organizations should make available new cloud services and cloud tools to their respective data scientists so that they can organize massive data quickly. Therefore, cloud tools are very important for a data scientist to analyze large amounts of data in a shorter period. It will save the company's time and help build strong and robust data models.

CHAPTER - 11
DATA MINING

Data mining is an elaborate process where the analyst searches for and identifies correlations, patterns, and anomalies within a given set of data. Based on their research, the findings are used for predictive analysis. There are several techniques that can be used to interpret the information obtained through data mining. In many environments, data mining reports help in making important revenue decisions, managing costs, risk mitigation, managing relationships with clients and other businesses, and so much more. In data science, data mining is also referred to as knowledge discovery.

Data mining involves the use of several tools from diverse fields, such as artificial intelligence and statistics. You also need some database management skills to help you learn how to analyze data sets better. Data mining is an important subject since the techniques used support other disciplines and applications. Recommendation models used in machine learning and search engine algorithms are some of the major areas where data mining is used frequently.

Types of Data Mining

Data mining is closely related to machine learning. Machine learning models collect data that is used for autonomous testing, such that in the long run, the models can make their own decisions. Data mining involves collecting this information for a specific purpose. From the machine learning relationship, we can look at data mining in terms of the learning processes. This gives us supervised and unsupervised learning. Let's look at this in-depth:

Supervised Learning

Supervised learning models are built for classification and prediction. It is an elaborate process that focuses on a predetermined output. In supervised learning, the goal of the learning model is to determine the value of any observation.

Spam filters, for example, determine whether your emails are legitimate or spam, and classify them accordingly. So first, the spam filters will predict the nature of each email, then send them to the spam folder instead of the inbox.

The following are some of the analytical models that are used in supervised data mining that you will learn over time:

- **Linear Regression**

Linear regression in data mining is a model that anticipates the value of continuous variables in a dataset, regardless of the number of data inputs. For example, in the real estate business, house agents use a number of factors to predict the value of a listing. Some of the factors they consider include the zip code, the year the house was built, the number of bedrooms, number of bathrooms, size of the compound, and other fixtures or amenities available.

- **Logistic Regression**

Logistic regression focuses on predicting the probability of a specific variable occurring. Like linear regression, this is also done regardless of the number of independent data inputs. An example of this in the application is in the financial lending sector. Lenders determine the creditworthiness of a borrower to establish whether they will default on loans using factors like their income, age, credit rating, and many other personal factors.

- **Regression Trees (Classification)**

Regression trees are used in data mining to predict outcomes for specific and continuous variables. Depending on the data available, these models are designed to generate specific rules that create clusters for variables that share similarities. The predicted value of an observation, therefore, becomes the group it is assigned to, based on the inherent features.

- **Time Series**

Time series analysis for data mining is about forecasting independent variables. These models can be used to predict how demand changes based on different factors. For example, in the retail sector, demand can be a function of time because customer demand changes with the seasons. With this information, management can make arrangements and restock their stores accordingly so that they do not suffer inventory issues.

- **Neural Networks**

These are analytical data models whose operation borrows a lot from the operations of the human brain. Neural networks are structured around the brain and the network of neurons. Neural networks in data mining depend on signals. This process is preferred in many organizations because it delivers results almost instantly. A good example of data mining in use through neural networks is in self-driving cars. These cars are trained to read traffic and other stimuli in their immediate environment so that they can make split-second decisions on how to get from one point to the other.

Unsupervised Learning

Unlike supervised learning models, unsupervised learning tasks are modeled to understand and describe data in a manner such that they can identify patterns and trends within the data. Among the best examples of this are recommendation systems. Since users are diverse, these models track user patterns, identifying personalized recommendations that can be used by decision-makers to improve the experience that customers have.

The following are some of the analytical models that are used in unsupervised data mining that you will learn over time:

- **Clustering**

Clustering models are built to identify similar data and group them together. These models are ideal in situations where you are using complex data to define a single outcome.

- **Principal Component Analysis**

In many cases, you come across datasets where the correlation between the new variables and input variables is not explicitly obvious. This is where the principal component analysis comes in handy. What happens here is that the model will record the same information from the input variables and new variables, but not with the same variables. The idea here is to enhance the accuracy and utility of data used in supervised data mining by reducing the variables under consideration.

- **Association Analysis**

This model is used to determine items that can be paired together. It is also referred to as market basket analysis. One of the best places where you will find this in action is at the supermarket. In many cases, they pair some items together and spread them evenly within the store. The idea is to make sure that you can see and purchase more items, in the process, increasing their sales volume.

Data Mining Applications

Data mining takes place in virtually all industries at the moment. Data is at the heart of most decisions that businesses make, and to gain all the data they need, it is important that they invest in some form of data mining. The depth of the process is irrelevant, as long as they obtain the data they need to suit their needs, and at the same time meet the regulations set by different organizations and conventions that they are party to such as the GDPR, which outline procedures, processes, and requirements for data handling.

The following are some of the instances where data mining takes place in different organizations and industries around us:

- **Spam Filtration**

We receive a lot of emails today, most of which are harmful in one way or the other. Spam mail is a headache for everyone. All the primary email service providers we use have spam filtration techniques in place to protect users from phishing attacks, malware, and any other issues that might arise from such mail.

Using data mining techniques, their systems monitor all emails in the process learning the characteristics of malicious emails. This report is then used to alert different security apparatuses in place, and mark the emails as spam. This is one aspect of machine learning that you will learn at an advanced stage, and you can actually build your own spam filters. Other than detecting spam messages, there are mechanisms in place that will go the extra mile and filter the messages to the spam folder before they get to the user's inbox.

- **Bioinformatics in the Healthcare Industry**

Within the healthcare industry, data mining helps medical experts determine the possibility that a patient might be at risk of certain diseases. This information is obtained from a database of the common risk factors. Other than that, they also look at the genetic makeup, family history, and demographics, which are modeled around unique features such that it is possible to identify certain health concerns before they turn into a full-blown outbreak.

- **Anti-Fraud Mechanisms**

The financial market is full of opportunities for fraudsters. From users who do not sign out of their accounts to those who use free public internet services and shared devices to access important details about their accounts, many users are vulnerable out there. As a result, financial institutions have taken it upon themselves to build and deploy data mining systems that can easily detect and prevent illegal transactions.

Computer forensics is one department in finance that is gaining popularity over time. This is because fraudsters are also benefiting from technological advancements and getting smarter by the day. The kind of data mining that takes place in financial institutions is stealthy, such that customers enjoy incredible protection without knowing about it.

One of the techniques that they use is to monitor customer spending habits. With this information, the fraud detection and prevention models are alerted in the event of an anomaly in the transaction behaviors. Some models withhold payments or block transactions until the customer can verify the transactions.

- **Recommendation Programs**

Recommendation systems are very popular today in online retail. Each time you visit an online retailer, you find recommendations popping up as you navigate their page. Other than that, when you view a product online, you receive recommendations on other products that would be ideal purchases alongside the one you are checking out.

When recommendation systems were introduced, they were not as effective as they are today. However, over time these systems have been integrated with machine learning models to make them more accurate and suitable to your needs. This is possible through predictive consumer behavior.

The machine learning models in use often monitor the behavior of customers all over the world. Using unique algorithms, they are able to create clusters of different customers based on their purchase habits. Today, some of the leading retailers in the industry do not just use data mining models, they build their own models that address their needs accordingly. This is how Amazon has managed to become a leader in online retail.

Data mining allows retailers to understand customer experiences better. Through this data, they are able to improve customer experiences, and this is one of the reasons why they have loyal customers who will purchase from them all the time. Another example of a company that has used recommendation systems effectively is Netflix. The need for an accurate recommendation system was so important to their business model that they offered a million dollars to developers who would build an algorithm that improved their recommendation system. Eventually, they got a system that delivered an 8% improvement. Over the years, they have refined the model to deliver even more accurate results.

- **Credit Scores and Risk Management**

Each time you want to borrow money from a bank or any other financial institution, they must determine your creditworthiness first. This helps the lender determine whether you are a risky borrower and, based on your risk profile, the interest rate to charge you if they agree to extend the loan facility to you.

CONCLUSION

It is a great time to be a data scientist! With each passing day, there is always something new happening in the field. Are you looking to become a data scientist soon? Well, there are some things which you can do to make sure you are absolutely ready for what lies ahead. Yes, this has nothing to do with more codes. Relax. Here are some of the tips from us to ensure that you have a wonderful career ahead of you.

Plan

You probably feel very comfortable taking on some forms of data science after finishing this book. If you feel that excitement, then you are on the right track. However, there is still a need to plan well before delving into it.

There are just many things which you will need to understand before you can achieve your goals, and without the right plan, this will be virtually impossible. So make sure you get yourself a checklist before taking the leap. Trust me, it will be important in the long run.

Planning will also ensure that you know just what aspects of data science you like the most. Knowing this will ensure that you know what you are interested in as well as your strengths and your weaknesses heading into the newfound career.

Read more

In this world of data science, reading will continue to be one important way of making sure that you keep advancing. Thanks to the internet, there are so many books which can help you to achieve your aim. So make sure that you are always reading and exploring new things and concepts. The only constant thing in the data science industry is that things are constantly changing with each passing day. So make sure that you stay on top of things.

Seek out the professional community

Fortunately for you, there are a lot of experts out there who can be considered experts in all matters relating to data mining. While this book gives you quite the beginners' guide, as you go deeper, you will need to gather some more experience. There is only so much experience which you can get from reading books. You will eventually have to seek out the experts in the field. They provide some valuable tips, which you will probably not find in any books. So there.

Practice every day

You cannot become a great data scientist without actually putting in the work. You will need to work to develop yourself in the field every day. The truth is that with data science, there is always something new to learn, and you will surely be doing yourself a world of good by practicing. Do you know the best part about practicing? It gets easier! Soon, you will be used to it and will be geared to go on with each passing year.

Take the Leap

You're finally ready. Having cold feet? It is time for you to take the leap. You have already invested so much into learning the trade, and it is time to kick off your career in data science. What if you are the owner of an organization looking for ways to make sure that your company gets the best of data science and what it offers? Then. It is also time to take the leap and make sure your company keeps on reaping the fruits of the integration of data science into your business.

Get a job and keep networking

There is so much more to explore out there. Keep working and expand your horizons past data science. All you need is a measure of confidence. Once you have that, the world will be at your feet.

The next step is to start putting some of the information that we have tackled in this guidebook to good use. As a business, if you have not already started to collect data from various sources, whether online, social media, from the customers who shop on your site or more, then you are already falling behind. It is only once that information is collected that you can begin the real work of sorting through all of that data and figuring out some of the information that is hidden inside. This guidebook spent some time looking at this process, and all of the steps that you can take to make data science, with the help of the Python coding language, work for you.

This guidebook has provided us with a lot of different information on data science, on how it works, machine learning, the Python language, and even some of the examples of how you can put all of this together and actually make it all work. Often data science sounds difficult and too hard to work on, but this guidebook has shown us some of the practical steps that we can take to put it all together.

When you are finally ready to take on some of the data that you have been accumulating, and you are excited to make this all work for you in terms of providing better customer service, and really seeing some good results in the decisions that you make for your business, make sure to check out this guidebook to help you get started with Python for data science.

Finally, you are right in the end. Keep your head up. Being focused and learning more about data science will turn out to be the best decision you will have made this year.

CPSIA information can be obtained
at www.ICGtesting.com
Printed in the USA
LVHW081515120621
690063LV00003B/86